To Emma and Silky   D.L.

To the one I love   P.V.

Published by
The Dial Press
1 Dag Hammarskjold Plaza
New York, New York 10017

First published in Great Britain by
Methuen Children's Books Ltd.
in association with Walker Books.

Library of Congress Cataloging in Publication Data
Lloyd, David. Air.
Summary: Introduces the various properties
and behavior of air, its composition,
and its importance to life on earth.
1. Air—Juvenile literature—Pictorial works. [1. Air]
I. Visscher, Peter, ill. II. Title.
QC161.2.L56 1983   551.5   82-9440
ISBN 0-8037-0141-1   AACR2

AIR

by DAVID LLOYD
Pictures by PETER VISSCHER

THE DIAL PRESS/New York

Wherever you live, you are surrounded by air. Even though you cannot see it or catch a handful of it, it is always around you, in every empty space. All day and night you breathe air, and the invisible gases from which it is made help to keep you alive. You can survive for weeks without food and for days without water, but without air you would die in minutes. Without air there would be no life.

Plants purify the air by releasing a gas called oxygen. Oxygen is what all human beings and animals breathe; it is what makes life possible. In addition, plants take in carbon dioxide, which we exhale. Mixed up with the oxygen and carbon dioxide in air are other gases, mostly nitrogen and some water vapor. No matter what the weather, there is always water in the air. Clouds are formed when water vapor condenses into tiny droplets that will fall to the earth as rain.

When the wind blows, it is air on the move. Wind shows the strength of air. Mighty rivers of air flow regularly around the world, governing the weather. Trees shake and bend to let it pass; leaves and seeds are scattered by it. Sycamore seeds spin away like helicopters, and dandelion seeds float away like miniature parachutes. The power of air—as wind—helps many plants to spread and spring up in new places.

Swiftly moving air can make the wildest winds of all. Hurricanes whirl across the land, wrecking everything in their path. While lightning flickers above, a tornado whips trees, parts of houses, chickens, horses, cars, even people, up into the sky. The air forms a roaring funnel, spinning faster than any other wind on earth, sucking things up or smashing them as it races past.

For millions of years, birds, bats, and insects have been masters of the air. Many of a bird's bones are built like a honeycomb, inside, with hollow spaces, to combine strength with lightness. Huge breast muscles power its wings. From earliest times, people dreamed of flying, of swooping through the air like birds. They constructed feathered wings and attached them to their arms. But their bodies were always too heavy, and the muscles of their arms too weak, so they never stayed off the ground for long.

The first people to fly successfully did so in a hot-air balloon over Paris in 1783. Because hot air weighs less than cool air, it always rises. Heat the air in a balloon, and the balloon also rises. Attach a basket underneath, climb aboard, and you will be lifted with it. High above the earth, you can hear only the creaking of the basket and faint sounds from far below. You become part of the wind, going where the wind goes.

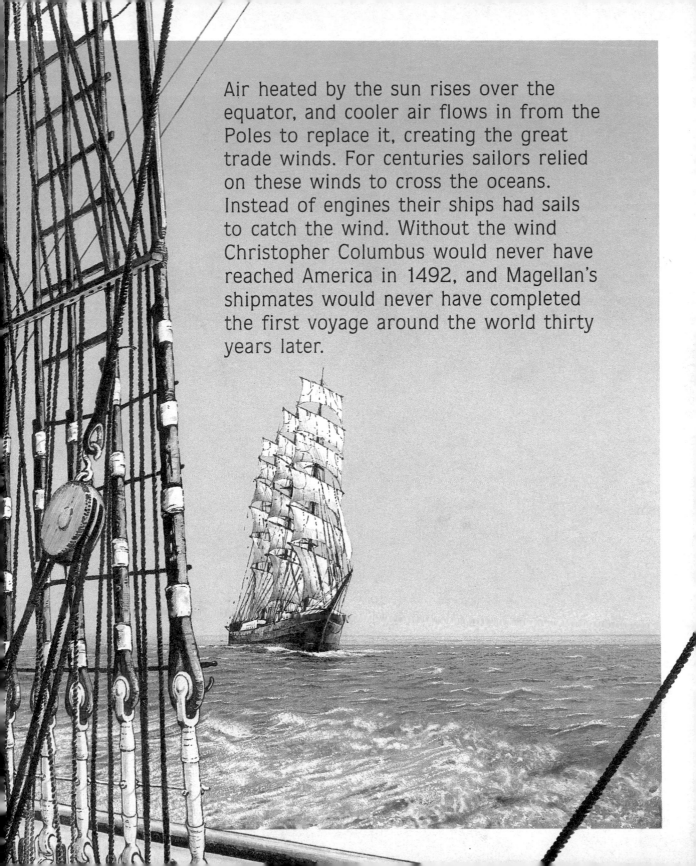

Air heated by the sun rises over the equator, and cooler air flows in from the Poles to replace it, creating the great trade winds. For centuries sailors relied on these winds to cross the oceans. Instead of engines their ships had sails to catch the wind. Without the wind Christopher Columbus would never have reached America in 1492, and Magellan's shipmates would never have completed the first voyage around the world thirty years later.

On land, wind has also been put to many uses. At one time windmills were widely used to grind corn and drain or irrigate the land. As the wind turned the sails of the mill, the machinery that was connected inside was set in motion. Sawmills were also powered in this way. In a few flat windswept places, the old sails still turn, harnessing the power of air.

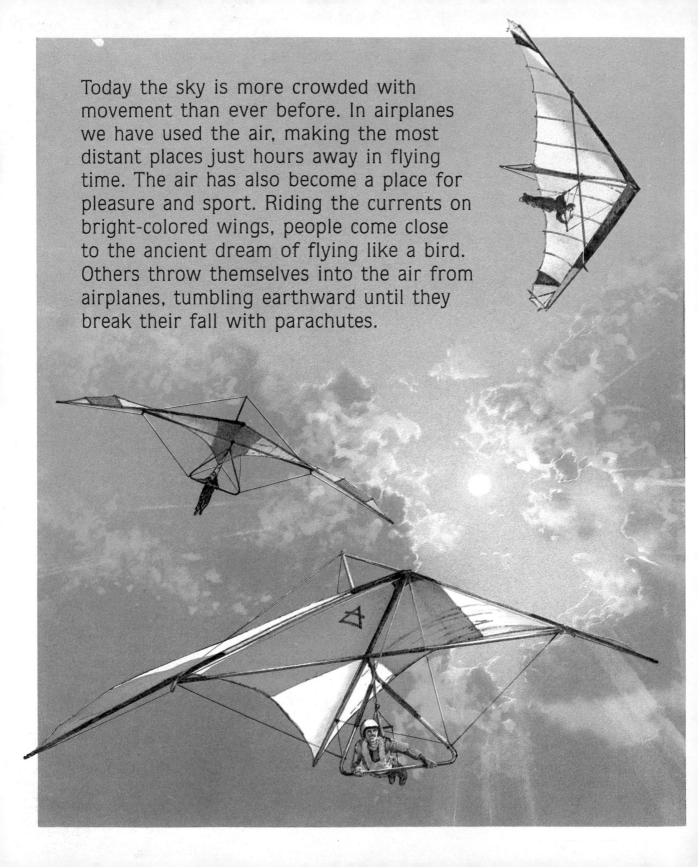

Today the sky is more crowded with movement than ever before. In airplanes we have used the air, making the most distant places just hours away in flying time. The air has also become a place for pleasure and sport. Riding the currents on bright-colored wings, people come close to the ancient dream of flying like a bird. Others throw themselves into the air from airplanes, tumbling earthward until they break their fall with parachutes.

But air can get dirty. The enormous activity and movement in big cities can destroy its freshness. Fumes and smells can invade it, making it unhealthful and difficult to breathe. Plants, which purify air naturally, can be poisoned when it is foul. Exhaust from cars and airplanes, smoke from chimneys—these can spoil the air. Air is something to look after carefully. It is best when it is clean.

As you go higher, the blanket of air covering the world, the atmosphere, becomes thinner and lighter. On the tops of the highest mountains there is hardly any air at all. Here a climber grows tired from lack of air. He finds it harder and harder to move or climb higher. He could die, unless he has brought with him an extra supply of air, compressed into cylinders, on his back.

Many miles up in the sky, but before you reach the moon, the atmosphere ends. There is no air in space. Man travels above the earth with his own supply of air. As he looks back from the silent emptiness, he sees air swirling about his planet, creating wind, causing weather, making life possible. The man rolls over in space, alone in the safety of his space suit. Far below are the people, plants, and animals of earth, all living in the air.

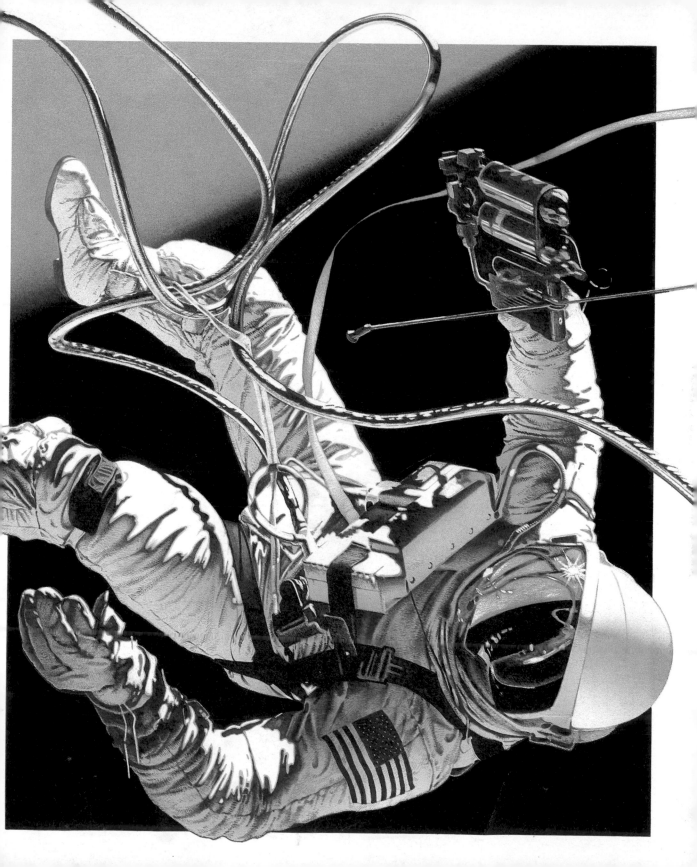